**BN-12** *The European Approach to Worker-Management Relationships,* by Innis Macbeath, October 1973 (£1.00, $2.50)

**BN-11** *An International Grain Reserve Policy,* by Timothy Josling, July 1973 (40p, $1.00)

**BN-10** *Man and His Environment,* by Harry G. Johnson, August 1973 (40p, $1.00)

**BN-9** *Prospective Changes in the World Trade and Monetary System: A Comment,* A Statement by the BNAC, October 1972 (30p, $.75)

**BN-8** *Multinational Corporations in Developed Countries: A Review of Recent Research and Policy Thinking,* by Sperry Lea and Simon Webley, March 1973 (80p, $2.00)

**BN-7** *Sterling, European Monetary Unification, and the International Monetary System,* by Richard N. Cooper, March 1972 (40p, $1.00)

**BN-6** *Multinational Corporations and British Labour: A Review of Attitudes and Responses,* by John Gennard, January 1972 (80p, $2.00)

**BN-5** *The Strategic and Political Issues Facing America, Britain and Canada,* by Leonard Beaton, October 1971 (40p, $1.00)

**BN-4** *Purposes and Prospects,* A Policy Statement describing the BNAC, April 1971 (20p, $.50)

**BN-3** *British Entry to the European Community: Implications for British and North American Agriculture,* by John S. Marsh, together with *Agriculture Policies for World Trade Expansion,* A Statement by the BNAC, March 1971 (50p, $1.25)

**BN-2** *Transatlantic Relations in the Prospect of an Enlarged European Community,* by Theodore Geiger, November 1970 (60p, $1.50)

**BN-1** *An Overall View of International Economic Questions Facing Britain, the United States and Canada during the 1970s,* by Harry G. Johnson, June 1970 (40p, $1.00)

---

**Occasional Paper-3** *Managing Product Innovation,* The views of five BNAC members: Carrol Dolen, Ken Durham, Walter Light, Chester Sadlow and Viscount Weir, October 1984 (£1.50 US$2.00, CAN$2.50)

**Occasional Paper-2** *Managing the Response to Industrial Decline,* by Joseph L. Bower, September 1984 (£1.50, US$2.00, CAN$2.50)

**Occasional Paper-1** *New Investment in Basic Industries,* prepared by a Committee Task Force, June 1979 (60p, $1.00)

*Publications of the British-North American Committee are available from:*

**In Great Britain and Europe**
BRITISH-NORTH AMERICAN RESEARCH ASSOCIATION
1 Gough Square, London EC4A 3DE
Telephone: 01-353 6371

**In the United States of America**
NATIONAL PLANNING ASSOCIATION
1606 New Hampshire Avenue, NW
Washington, DC 20009
Telephone: 202-265-7685

**In Canada**
C. D. HOWE INSTITUTE
2275 Bayview Avenue, Toronto, Ontario M4N 3M6
Telephone: 416-485-4310

# Publications of the British-North American Committee

**BN-34** *The Process of Innovation,* by Nuala Swords-Isherwood, October 1984 (£5.00, US$8.00, CAN$10.00)

**BN-33** *Governments and Multinationals: Policies in the Developed Countries,* by A. E. Safarian, December 1983 (£4.00, $8.00)

**BN-32** *Trade Issues in the Mid 1980s,* by Sidney Golt and a Committee Policy Statement, October 1982 (£3.50, $7.00)

**BN31** *The Newly Industrializing Countries: Adjusting to Success,* by Neil McMullen, November 1982 (£3.50, $7.00)

**BN-30** *Conflicts of National Laws with International Business Activity: Issues of Extraterritoriality,* by A. H. Hermann, August 1982 (£3.00, $6.00)

**BN-29** *Industrial Innovation in the United Kingdom, Canada and the United States,* by Kerry Schott, July 1981 (£2.25, $5.00)

**BN-28** *Flexible Exchange Rates and International Business,* by John M. Blin, Stuart I. Greenbaum and Donald P. Jacobs, December 1981 (£3.00, $8.00)

**BN-27** *A Trade Union View of US Manpower Policy,* by William W. Winpisinger, April 1980 (£1.75, $3.00)

**BN-26** *A Positive Approach to the International Economic Order, Part II: Non-trade Issues,* by Alasdair MacBean and V. N. Balasubramanyam, May 1980 (£2.25, $5.00)

**BN-25** *New Patterns of World Mineral Development,* by Raymond F. Mikesell, September 1979 (£2.25, $5.00)

**BN-24** *Inflation is a Social Malady,* by Carl Beigie, March 1979 (£2.00, $4.00)

**BN-23** *A Positive Approach to the International Economic Order, Part I: Trade & Structural Adjustment,* by Alasdair MacBean, October 1978 (£1.75, $3.00)

**BN-22** *The GATT Negotiations 1973-79: The Closing Stage,* by Sidney Golt and a Committee Policy Statement, May 1978 (£1.50, $3.00)

**BN-21** *Skilled Labour Supply Imbalances: The Canadian Experience,* by William Dodge, November 1977 (£1.50, $3.00)

**BN-20** *The Soviet Impact on World Grain Trade,* by D. Gale Johnson, May 1977 (£1.75, $3.00)

**BN-19** *Mineral Development in the Eighties: Prospects and Problems,* a Report Prepared by a Group of Committee Members with a Statistical Annex by Sperry Lea, November 1976 (£1.50, $3.00)

**BN-18** *Skilled Labour Shortages in the United Kingdom: With Particular Reference to the Engineering Industry,* by Gerry Eastwood, October 1976 (£1.50, $3.00)

**BN-17** *Higher Oil Prices: Worldwide Financial Implications,* a Policy Statement by the British-North American Committee and a Research Report by Sperry Lea, October 1975 (£1.50, $3.00)

**BN-16** *Completing the GATT: Toward New International Rules to Govern Export Controls,* by Fred Bergsten, October 1974 (80p, $2.00)

**BN-15** *Foreign Direct Investment in the United States: Opportunities and Impediments,* by Simon Webley, September 1974 (80p, $2.00)

**BN-14** *The GATT Negotiations, 1973-75: A Guide to the Issues,* by Sidney Golt, April 1974 (£1.00, $2.00)

**BN-13** *Problems of Economic Development in the Caribbean,* by David Powell, compiled from a study by Irene Hawkins, November 1973 (80p, $2.00)

**The C. D. Howe Institute** was established in 1973 by the merger of the C. D. Howe Memorial Foundation and the Private Planning Association of Canada. It is a nonprofit, nonpolitical organization seeking to contribute nonpartisan research findings and commentary on Canadian economic policy issues.

The guiding principle of the Institute is to conduct its research and analysis in a manner that is balanced in approach, professional in method and readable in style.

To ensure diversity in perspective, participation is encouraged from organized labor, business, agricultural associations, and the professions.

While its focus is national, the Institute recognizes that Canada is composed of regions, each of which may have a particular point of view on policy issues, unique interests and concerns, and different concepts of national priorities. The Institute also pursues involvement from both of Canada's major linguistic communities.

It is not the purpose of the Institute to promote consensus on policy issues, although on occasion that may be feasible. The primary function is to add to public understanding of issues by providing sound analysis reflecting objective treatment of diverse points of view. The staff of the Institute seeks to develop good working relationships with public officials for the purpose of better understanding the basis for government decisions and contributing effectively to public policy formulation.

A Board of Directors is responsible for the general administration of the Institute, while the immediate direction of the policies, program and staff in vested in the Executive Director. The function of the Board is to make independent research and publication possible under the most favourable conditions and not to control the conduct and conclusions of research activity.

Mr. Michael Belanger is Chairman, Dr. Wendy Dobson is Executive Director and Treasurer, and Ms. Wendy Walker is Corporate Secretary.

The Institute's publications, including those of the British-North American Committee, are available from its offices: Suite 900, 555 Dorchester Blvd., Montreal, Quebec H2Z 1B1; P.O. Box 1621, Calgary, Alberta T2P 2L7; and Glendon Hall, 2275 Bayview Avenue, Toronto, Ontario M4N 3M6 (Tel. 416-485-4310).

# Sponsoring Organizations

**The British-North American Research Association** was inaugurated in December 1969. Its primary purpose is to sponsor research on British-North American economic relations in association with the British-North American Committee. Publications of the British-North American Research Association as well as publications of the British-North American Committee are available from the Association's office, 1 Gough Square, London EC4A 3DE (Tel. 01-353 6371). The Association is recognized as a charity and is governed by a Council under the chairmanship of A. B. Marshall.

**The National Planning Association** is an independent, private, nonprofit, nonpolitical organization that carries on research and policy formulation in the public interest. NPA was founded during the Great Depression of the 1930s when conflicts among the major economic groups—business, labor, agriculture—threatened to paralyze national decision making on the critical issues confronting American society. It was dedicated to the task of getting these diverse groups to work together to narrow areas of controversy and broaden areas of agreement and to provide on specific problems concrete programs for action planned in the best traditions of a functioning democracy. Such democratic planning, NPA believes, involves the development of effective governmental and private policies and programs not only by official agencies but also through the independent initiative and cooperation of the main private-sector groups concerned. And to preserve and strengthen American political and economic democracy, the necessary government actions have to be consistent with, and stimulate the support of, a dynamic private sector.

NPA brings together influential and knowledgeable leaders from business, labor, agriculture, and the applied and academic professions to serve on policy committees. These committees identify emerging problems confronting the nation at home and abroad and seek to develop and agree upon policies and programs for coping with them. The research and writing for these committees are provided by NPA's professional staff and, as required, by outside experts.

In addition, NPA's professional staff undertakes research designed to provide data and ideas for policymakers and planners in government and the private sector. These activities include the preparation on a regular basis of economic and demographic projections for the national economy, regions, states, metropolitan areas, and counties; research on national goals and priorities, productivity and economic growth, welfare and dependency problems, employment and manpower needs, and energy and environmental questions; analyses and forecasts of changing international realities and their implications for U.S. policies; and analyses of important new economic, social and political realities confronting American society.

NPA publications, including those of the British-North American Committee, can be obtained from the Association's office, 1606 New Hampshire Avenue, N.W., Washington, D.C. 20009 (Tel. 202-265-7685).

JAMES R. SCHLESINGER
Senior Advisor, Lehman Brothers,
Sherson/Lehman/American Express Inc.,
New York, N.Y.

LORD SEEBOHM
Dedham, Essex

THE EARL OF SELKIRK
President, Royal Central Asian Society,
London

JACK SHEINKMAN
Secretary-Treasurer, Amalgamated
Clothing & Textile Workers' Union, New
York, NY

LORD SHERFIELD
Chancellor of Reading University,
London

R. MICHAEL SHIELDS
Managing Director, Associated
Newspapers Group, plc., London

GEORGE L. SHINN
Chairman and Chief Executive Officer
Emeritus, The First Boston Corporation,
New York, NY

GORDON R. SIMPSON
Chairman, General Accident Fire and Life
Assurance Corporation plc., Perth,
Scotland

SIR ROY SISSON
Chairman, Smiths Industries plc., London

SIR LESLIE SMITH
Chairman, BOC International plc., London

RALPH I. STRAUS
New York, NY

THOMAS H. B. SYMONS
Vanier Professor, Trent University, and
Director, Celanese Canada, Inc.,
Peterborough, Ontario

ALLAN R. TAYLOR
President and Chief Operating Officer,
The Royal Bank of Canada, Toronto,
Ontario

JAMES O. THACKRAY
Chairman, Bell Canada, Montreal,
Quebec

W. C. THOMSON
A Managing Director, Royal Dutch/Shell
Group of Companies, London

ALEXANDER C. TOMLINSON
President, National Planning Association,
Washington, DC

ALAN TUFFIN
General Secretary, Union of
Communication Workers, London

J. C. TURNER
General President, International Union of
Operating Engineers, AFL-CIO,
Washington, DC

W. O. TWAITS
Toronto, Ontario

JOHN A URQUHART
Executive Vice President and Sector
Executive, International Sector, General
Electric Company, Fairfield, Connecticut

MARTHA REDFIELD WALLACE
President, Redfield Associates, New
York, NY

GLENN E. WATTS
President, Communications Workers of
America, AFL-CIO, Washington, DC

VISCOUNT WEIR
Chairman, The Weir Group Limited,
Glasgow, Scotland

FREDERICK B. WHITTEMORE
Managing Director, Morgan Stanley &
Co. Incorporated, New York, NY

MARGARET S. WILSON
Chairman of the Board, Scarbroughs,
Austin, Texas

CHARLES WOOTTON
Director, International Public Affairs, Gulf
Oil Corporation, Pittsburgh, Pennsylvania

## Members of the British North American Committee

WALTER F. LIGHT
Chairman and Chief Executive Officer,
Northern Telecom Ltd., Mississauga,
Ontario

FRANKLIN A. LINDSAY
Chairman, Engenics Corporation,
Lincoln, Massachusetts

SIR PETER MACADAM
Former Chairman, B.A.T. Industries plc.,
London

IAN MacGREGOR
Honorary Chairman, AMAX Inc.,
Greenwich, Connecticut

CARGILL MacMILLAN, JR.
Senior Vice President, Cargill Inc.,
Minneapolis, Minnesota

JOHN N. MALTBY
Chairman, Burmah Oil plc., Swindon

J. P. MANN
Deputy Chairman, United Biscuits
(Holdings) Ltd., Middlesex

PAUL E. MARTIN
President and Chief Executive Officer,
The CSL Group, Inc., Montreal, Quebec

WILLIAM J. McDONOUGH
Executive Vice President and Chief
Financial Officer, The First National Bank
of Chicago, Chicago, Illinois

DONALD K. McIVOR
Chairman and Chief Executive Officer,
Imperial Oil Limited, Toronto, Ontario

DONALD E. MEADS
Chairman and President, Carver
Associates, Plymouth Meeting,
Pennsylvania

SIR PATRICK MEANEY
Chairman, The Rank Organisation plc.,
London

JOHN MILLER
Vice Chairman, National Planning
Association, Washington, DC

WILLIAM R. MILLER
President, Pharmaceutical & Nutritional
Group, Bristol-Myers Company, New
York, NY

DAVID MORTON
President and Chief Executive Officer,
Aluminium Co. of Canada Ltd., Montreal,
Quebec

JOHN J. MURPHY
Chairman, President & Chief Executive
Officer, Dresser Industries, Dallas, Texas

KENNETH D. NADEN
President, Volunteers in Overseas
Cooperative Assistance, Washington, DC

TED NEWALL
Chairman, President and Chief Executive
Officer, Du Pont Canada Inc.,
Mississauga, Ontario

WILLIAM S. OGDEN
Chairman, Continental Illinois National
Bank and Trust Company of Chicago,
Chicago, Illinois

ANTHONY J. E. O'REILLY
President and Chief Executive Officer,
H. J. Heinz Company, Pittsburgh,
Pennsylvania

LORD PENNOCK
Chairman, BICC plc., London

D. A. S. PLASTOW
Managing Director, Vickers plc., London

GEORGE J. POULIN
General Vice President, International
Association of Machinists & Aerospace
Workers, Washington, DC

SIR RICHARD POWELL
Hill Samuel Group Limited, London

ALFRED POWIS
Chairman and Chief Executive Officer,
Noranda Mines Limited, Toronto, Ontario

PAUL E. PRICE
Executive Vice-President, International
Grocery Products, The Quaker Oats
Company, Chicago, Illinois

GRANT L. REUBER
President, Bank of Montreal, Toronto,
Ontario

BEN ROBERTS
Professor of Industrial Relations, London
School of Economics, London

HAROLD B. ROSE
Group Economic Adviser, Barclays Bank
plc., London

CHESTER A, SADLOW
Executive Vice-President, Advanced
Production Technology, Westinghouse
Electric Corp., Pittsburgh, Pennsylvania

WILLIAM R. SALOMON
Honorary Chairman, Salomon Brothers,
Inc., New York, NY

A. C. I. SAMUEL
Handcross, Sussex

HOWARD D. SAMUEL
President, Industrial Union Department,
AFL-CIO, Washington, DC

NATHANIEL SAMUELS
Advisory Director, Lehman Brothers
Kuhn Loeb Inc., New York, NY

THOMAS R. SAYLOR
Vice-President, Garnac Grain Co., Inc.,
New York, NY

EDWARD DONLEY
Chairman, Air Products and Chemicals, Inc., Allentown, Pennsylvania

ALLAN R. DRAGONE
Vice-Chairman, Burlington Industries, Inc., New York, NY

JOHN DU CANE
Director, AMAX Inc., London

SIR WILLIAM DUNCAN
Chairman, Rolls Royce Limited, London

KEN DURHAM
Chairman, Unilever plc., London

GERRY EASTWOOD
General Secretary, Association of Patternmakers and Allied Craftsmen, London

HARRY E. EKBLOM
Chairman, University Science Partners Inc., Ridgewood, NJ

GLENN FLATEN
President, Canadian Federation of Agriculture, Regina, Saskatchewan

C. S. FLENNIKEN
Chairman, President and Chief Executive Officer, CIP Inc., Montreal, Quebec

DAVID W. FOX
Vice Chairman, The Northern Trust Co., Chicago

SIR ALISTAIR FRAME
Deputy Chairman and Chief Executive, Rio-Tinto Zinc Corporation, London

SIR CAMPBELL FRASER
President, Dunlop Holdings plc., London

ROBERT R. FREDERICK
President, RCA Corporation, New York, NY

GWAIN H. GILLESPIE
Executive Vice-President, Finance and Administration, Chief Financial Officer, R. J. Reynolds Industries, Inc., Winston Salem, NC

MALCOLM GLENN
Executive Vice-President, Reed Holdings Incorporated, Rickmansworth, Herts

PETER A. GRAHAM
Senior Deputy Chairman, Standard Chartered Bank plc., London

JOHN H. HALE
Managing Director, S. Pearson & Son plc., London

JOHN C. HALEY
Executive Vice-President, The Chase Manhatten Bank, New York, NY

HON. HENRY HANKEY
British Secretary, BNAC, Westerham, Kent

AUGUSTIN S. HART, JR.
Retired Vice-Chairman, Quaker Oats Company, Chicago, Illinois

FRED L. HARTLEY
Chairman and President, Union Oil Company of California, Los Angeles, California

H. J. HAYNES
Senior Counselor, Bechtel Group, Inc., San Francisco, California

PAUL HAZEN
Vice Chairman, Wells Fargo & Company, San Francisco, California

G. R. HEFFERNAN
President, Co-Steel International Ltd., Whitby, Ontario

ROBERT HENDERSON
Chairman, Kleinwort Benson, Lonsdale plc., London

PAUL J. HOENMANS
President, Marketing & Refining Division, Mobil Oil Corp., New York, NY

SIR TREVOR HOLDSWORTH
Chairman, Guest Keen & Nettlefolds plc., London

HENDRIK S. HOUTHAKKER
Professor of Economics, Harvard University, Cambridge, Massachusetts

DONALD P. JACOBS
Dean, J. L. Kellogg Graduate School of Management, Northwestern University, Evanston, Illinois

GEORGE S. JOHNSTON
President, Scudder, Stevens & Clark, New York, NY

PAUL J. KOFMEHL
Director General, Operations, IBM Europe, Paris

C. CALVERT KNUDSEN
Vice Chairman and Chairman, Executive Committee, MacMillan, Bloedel Limited, Vancouver, B.C.

GAVIN LAIRD
General Secretary, Amalgamated Union of Engineering Workers (Engineering Section), London

PETER LAISTER
Chairman and Chief Executive, Thorn EMI, London

H. U. A. LAMBERT
Chairman, Barclays Bank UK Ltd., London

DR. INGRAM LENTON
Managing Director, Bowater Corporation plc., London

WILLIAM A. LIFFERS
Vice Chairman, American Cyanamid Company, Wayne, New Jersey

# Members of the British-North American Committee*

*Chairmen*

A. B. MARSHALL
Chairman, Commercial Union Assurance, London

WILLIAM L. WEARLY
Chairman (Retired), Ingersoll-Rand Company, Woodcliff Lake, New Jersey

*Chairman, Executive Committee*

WILLIAM I. M. TURNER, JR.
Chairman and Chief Executive Officer, Consolidated-Bathurst Inc., Montreal, Quebec

*Members*

ROBIN W. ADAM
Chairman, MEPC, London

DR. DAVID ATTERTON
Chairman, Foseco Minsep Ltd., Birmingham

CHARLES F. BAIRD
Chairman and Chief Executive Officer, INCO Limited, Toronto, Ontario

JOSEPH E. BAIRD
Los Angeles, California

PROFESSOR SIR JAMES BALL
Chairman, Legal & General Assurance Group plc., London

ROBERT A. BANDEEN
Chairman and President, Crown Life Insurance Company, Toronto, Ontario

NICHOLAS H. BARING
Managing Director, Baring Brothers Co., Ltd., London

SIR DONALD BARRON
Chairman, Midland Bank plc., London

LAURENT BEAUDOIN
Chairman and Chief Executive Officer, Bombardier Inc. Montreal, Quebec

MICHEL BELANGER
Chairman of the Board, President and Chief Executive Officer, National Bank of Canada, Montreal, Quebec

W. J. BENSON
Deputy Chairman, National Westminster Bank plc., London

C. FRED BERGSTEN
Director, Institute for International Economics, Washington, DC

ROGER BEXON
Deputy Chairman, The British Petroleum Company, plc., London

CARROL D. BOLEN
Vice-President, Pioneer Hi-Bred International Inc., Des Moines, Iowa

THORNTON F. BRADSHAW
Chairman of the Board and Chief Executive Officer, RCA Corporation, New York, NY

F. S. BURBIDGE
Chairman and Chief Executive Officer, Canadian-Pacific Limited, Montreal, Quebec

SIR RICHARD BUTLER
President, National Farmers' Union, London

VISCOUNT CALDECOTE
Chairman, Investors in Industry Group plc., London

SILAS S. CATHCART
Chairman, Illinois Tool Works Inc., Chicago, Illinois

GEORGE J. CLARK
Executive Vice President, Citibank, N.A., New York, NY

FRANK J. CONNOR
President, American Can Company, Greenwich, Connecticut

DONALD M. COX
Director and Senior Vice President, Exxon Corporation, New York, NY

JAMES E. CUNNINGHAM
Chairman and Chief Executive Officer, McDermott International, Inc., New Orleans, Louisiana

JAMES W. DAVANT
Retired Chairman, Paine Webber Incorporated, New York, NY

RALPH P. DAVIDSON
Chairman, Time Incorporated, New York, NY

WILLIAM DODGE
Ottawa, Ontario

WILLIAM H. DONALDSON
Chairman and Chief Executive, Donaldson Enterprises Inc., New York, NY

---

*\*Membership of the Committee does not necessarily signify agreement with all the contents of its publications*

Again, analysis led naturally to the solution which was to make the turbine elements in one part by precision casting methods so that complicated and costly machining was unnecessary. A modular design was also adopted so that we could simply put on the shaft as many elements as were needed to provided the power a particular well required.

This concept, incidentally, solved one marketing problem which was the requirement – always difficult to meet – of being able to offer a custom designed piece of equipment on short delivery, without carrying a lot of stocks.

In parallel with these problems on the technical side of the innovation process, there were two other real ones. How could we afford the whole development programme which was very costly for a firm of our size, and how could we ensure that the market, which was a rather conservative one, would accept so radical an innovation?

The financing problem was not made easier by the fact that the features of our new design were such that we had to go a very long way down the whole development route, including manufacture and design of prototypes, to be really sure that we were going to have a satisfactory solution.

We therefore decided to try and kill both birds with one stone and go to an oil company and ask for their financial support. We reasoned that the oil companies had much the most to gain financially if this innovation succeeded and, moreover, that if we could get the backing of an oil industry major it should be considerably easier to gain industrial acceptance for the product. We also felt that a backer would be much more inclined to lend us a well for testing the equipment.

In fact, we decided as a business decision – and innovation is as much a matter of business decision as having good R & D people – that if we could not have found an oil company who would back us we would abandon the concept.

Here I must tell you that B.P. unhesitatingly agreed to support us with a large part of the cost, and shortly after they had made this decision to back us then B.N.O.C and the Offshore Supplies Office agreed to share the cost. We therefore started the development programme with the funding secure, and with the confidence of knowing that potential customers were sponsoring the project. I should add that since the development phase was completed the O.S.O. have continued to give us enthusiastic support.

# WEIR HYDRAULIC-DRIVE DOWNHOLE PUMP SYSTEM

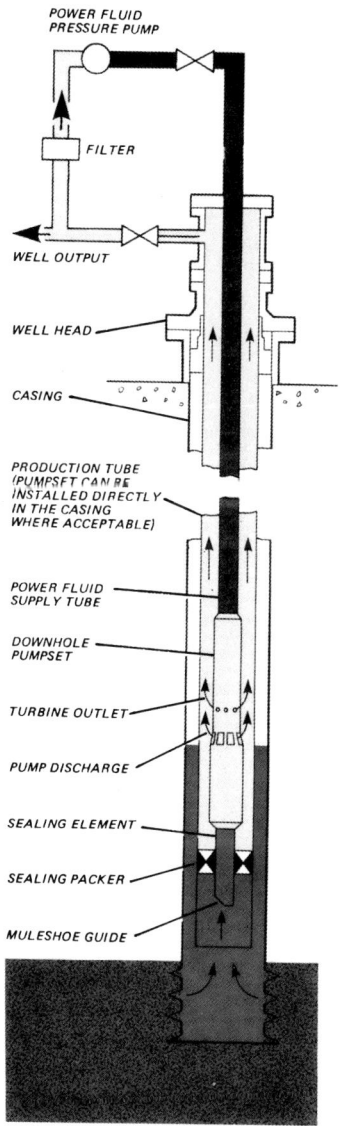

Obviously, at this point we had not done anything innovative at all. All we had done was to identify the problems; know that if we could solve them we would have an interesting market: and know that we needed a radical solution to solve them satisfactorily and to our profit.

**Causes of Problems**
The approach we then undertook to bridging the gap between the problem and the solution, was not to concentrate our thinking about finding a solution as such, but rather to decide what the problem really was and then analyse it.

Here I believe is another key to most innovative solutions. They do not very often emerge from "Eureka" type discoveries – or invention in the traditional sense of the word.

Instead it is the close analysis of the true nature of the indentifed problem which must often lead naturally to the innovation. Of course there are Eureka innovations of startling originality, but there is plenty of well documented history of the way in which many important innovations came about, and most of this suggest that they basically derived from the kind of analysis I have mentioned. James Watt's famous innovation with respect to the steam engine is a familiar one.

In the case of the Downhole pump we concluded that two real problems with the existing equipment were, first, that it was inherently unreliable because of its very long flimsy configuration and, second, that it was wrong in basic concept because it was driven by an electric motor under operating conditions which normal motor technology had difficulty in coping with.

It followed very obviously that the reason conventional machines were so long, was that they ran at a low speed, If you consider a pump impeller, the fact is that the pressure it will generate is proportional to the square of the speed at which it is driven.

If therefore we could drive the pump impellers much faster we could greatly reduce the length of the pump-end of the machine compared to a conventional unit. The obvious way to do this was to use a turbine and clearly therefore, a high speed hydraulic turbine-driven pump could – providing we could design it – provide the basis of a workable innovative conceptual solution.

Innovations however are not as simple as that. The next problem we faced was whether we could achieve all this without, so to speak, "shooting ourselves in the foot", by developing a conceptual solution which in practice produced its own particular problems in reliability and cost.

Once again, we used the same method of thinking as we had used to come up with the original conceptual solution. This was a thorough analysis of the potential problems.

One of the first problems was that to make our solution work practically and reliably required the use of very hard wearing material which it was difficult to machine economically. At the same time, almost every oilwell has different operating conditions so the design had to be an inherently flexible one.

market for machinery on our doorstep in Scotland, and although our company has long been the largest U.K. pump manufacturer it did not have a very strong product line for the oil industry.

We did have a good initial success in the North Sea with conventional pumping equipment adapted or designed to meet the requirements of this new market. Although new to us, these were in no sense innovatory products.

It was quickly clear to us that if we were to take worthwhile advantage of the local oil industry market, and then use the reference and experience gained in it to extend outwards and compete in wider international markets, we would have to adopt a more radical approach.

From an early point we had seen – as relative newcomers – that the oil equipment industry was a long established and mature one with high degree of conservatism among both users and suppliers. Our first constructive – if rather unoriginal – thought followed from those two propositions. It was simply the conclusion that it was fundamentally useless to try and develop products for such a market on a conventional "me too" basis, and just come up with much the same equipment as established and competent companies already supplied and compete with them.

Instead, we decided to look at a wide range of oilfield duties, searching systematically for instances where the established solution was inadequate to the real problem. This general approach is, we think, one of the real starting points for much innovative thinking. Downhole pumping was just such a case.

The established solution downhole for large wells is to use electro-submersible pumps. The problems of such pumps are considerable and I fear I must be a little technical in describing them.

First of all, the diameter of the driving motor is limited by the size of the casing in the oil-well, and if you have to turn out hundreds of horsepower from the motor, limitations to the power you can generate per foot length of a motor of a given diameter mean you end up with a very long motor. You also run into very difficult electrical insulation problems. You have further difficulties due to high temperatures, to corrosion and erosion damage, and serious technical problems with the cable to supply power to the motor. On top of all that, the limitations on the motor speed mean that you need a lot of impellers to generate the pressure required. Thus, the whole device becomes longer and longer and you can end up with a machine about 100ft. long and about 6ins. in diameter. This produces further problems if you want to deviate the oil well in a curve to get into the reservoir.

This all adds up to a basically unreliable product, particularly when you put it thousands of feet down the ground, and a product which oftens has to be removed from the well after only a few weeks' operation, simply because it has broken down.

It also adds up to a severe cost to the oilfield operator who may have to stop production while he pulls up the damaged pump and puts a new one down.

# An Innovation of the Weir Group

## by Lord Weir,
*Chairman of the Weir Group plc*

> *The Weir Group,* **founded as an engineering partnership in 1871, now employs about 4,600 people and manufactures pumps and associated equipment for the power generation, oil, marine, water supply and many other industries; steel castings and fabrications; and metal pattern equipment. It undertakes engineering contracting particularly in the supply of desalination plants. It has overseas subsidiaries or associates in Australia, Canada, France, Italy, India, the Middle East, the Netherlands, Singapore, and Spain.**

The purpose of this paper is to describe one of the recent innovations of the Weir Group in the field of pumping. I will start by saying something about the product, and will then try to explain the philosophy and the process of innovation which finally resulted in its production.

The particular product is a "Downhole pump" a diagram of which is shown in Figure 1. Some people may have seen this product on a B.B.C. television programme showing the finalists in the Prince of Wales' annual prize for innovation: it was the runner up out of 2,000 entries.

The pump is put down oil-wells when the natural pressure drops and the flow of oil needs to be boosted. Downhole pumps of one kind or another have been used for a long time now and the "nodding donkey" pump is a familiar sight in many oil-fields.

The innovatory concept of this particular pump is that it is driven by a high speed, high pressure, hydraulic turbine, and is one-tenth of the size of the product it is designed to replace.

The pump impellers are mounted on the same shaft as the turbine element and are driven by a number of these turbine stages mounted on top of each other on the shaft.

To drive the turbine, there is a high pressure pump on the surface and power fluid – either water or oil – is piped down the well through a small tube at high pressure. This powers the turbine; the turbine drives the pump; and up to the surface comes the oil production plus the power fluid which was put down in the first place.

It is such a very simple concept that it could reasonably be asked what is innovatory about it at all.

Rather than answer that now, I will instead start at the beginning of the whole process of innovation and hope to show why we ended up with this particular product – why the product is an innovatory one – and something about our philosophy of innovation.

## Markets
Our starting point was a marketing one. When North Sea oil came on the scene we immediately had open to us a large – if unfamiliar – new domestic

As Victor Hugo put it, there is no power on earth so strong as an idea whose time has come.

Clearly, the time has come for American, British and Canadian companies alike to renew their commitment to the innovation process. Our futures depend upon that commitment.

That takes strong management – management with conviction – crusaders who believe strongly enough in an idea to fight for it all the way to the marketplace.

It's as Joe Wilson, President of the former Haloid Corporation, said about his copier development – "in the final analysis, it was an act of faith." His company, by the way, is better known today as Xerox.

**Fifth,** big companies need to have more faith in innovation. To be fair to Westinghouse, and other big companies, we are conscious of these problems, and are striving to improve them. We have a program under way we like to refer to as our "*in*-trepreneurship" program in which we will be giving a small group of engineers a research idea that originated at our corporate R&D Center, and encouraging them to run with it as if they owned it. It's an exciting concept, for management and for the engineers involved and we think it is a model that can be replicated elsewhere.

Another development that shows a lot of promise in the U.S. is the establishment of new joint venture and shared research efforts with major universities. We are currently working with about half-a-dozen schools on some exciting projects, including robotics and artificial intelligence. Our effort with Carnegie-Mellon University has already paid off handsomely in terms of practical innovations. We have applied a number of C.M.U. ideas to our product line and automation problems within our own factory. Here again, the solution is putting the right people together in small groups.

The *sixth* point comes from the Japanese. In spite of public perceptions, and all the media hype, the Japanese are not the great innovators. The Japanese Ministry of Trade recently reported that over the past 20 years, there have been only 26 real technological innovations in Japan – only two of which could be described as "momentous". Meanwhile, in the U.S., there were 237 technical innovations, 65 of which have had great significance. The point is this: the Japanese know that they are imitators, and they are doing something about it. Hitachi, Mitsubishi and other companies are training their people in the innovation process to overcome this shortcoming. We too need to step up our educational efforts on the subject and establish innovation courses in engineering and business schools, and in-house training programs.

**Seventh,** as businessmen, scientists and engineers, we collectively share an awesome responsibility. If we take our jobs seriously, we must educate the public and assist in the formation of public policy on matters that impact innovation in our respective societies. No innovation will prosper unless our elected officials and the people they serve perceive its need as real.

**Finally,** in Western societies such as our own, the best innovations emerge from the private sector. But the private sector cannot do it alone. The public sector must address the larger issues of technical education and must provide adequate funding for large-scale innovations directed toward the common good of society.

Above all, to encourage innovation within our organizations, we must be innovators ourselves.

measure how well a company manages innovation. We need rewards for individuals who do it well.

Current financial reporting systems actually encourage managers *not* to be innovative. We have somehow to overcome this short-term, bottom-line mentality and develop an accounting system that takes into consideration the assets and liabilities of long-term research.

**Third,** we must do what we can to cultivate entrepreneurialism within large corporations. Small companies generate two-thirds of the patentable ideas filed in the U.S. each year. What that says is that small teams of engineers, when left alone to do their thing, are better at coming up with new ideas than large companies.

Studies have shown that a group of four to six people with complementary talents is the optimum size for developing innovative ideas. Anything over ten bogs down the creative juices. I'm sure you all read about how IBM, that giant of giants, developed its PC computer in such a short time. They gave a small group of engineers unprecedented autonomy and let them innovate to their hearts' content in a small building away from the mainstream of corporate activity.

My **fourth** bit of advice is a corollary to number three. In addition to allowing engineers to operate like entrepreneurs, managers must also be given that freedom. Today many large companies in the U.S. simply lack adequate entrepreneurial talent. Their management teams are populated with veteran "company men" whose visions are restricted to what the corporate policy manual says.

To survive as a large business today, you need astute corporate administrators *and* entrepreneurs. Certainly, as an industry grows, the entrepreneurial organization must have the support of the larger corporation and its resources.

Our Unimation subsidiary is an excellent example of this marriage of corporate and entrepreneurial strengths. Both parties are benefiting today, and we are learning a lot about the innovation process from each other.

Joe Engleberger, the founder of this robot manufacturing operation, is now working with us as a consultant – a role that he relishes. He is looking at entrepreneurial opportunities right inside our own company. We do a lot of new product development within Westinghouse, but we are not always sure we are getting the proper return on our investment. Joe has gone out into our plants and spoken to many of our middle management people. He has discovered that often, when an engineer has a new idea, he presents it to management, only to discover that, instead of encouragement, he hears dozens of reasons why it won't work.

Corporate people too often become steeped in regulatory proceedings, marketing studies, financial analyses and the like, that can quickly become the kiss of death for innovation. Selling an idea in a large company can be like throwing a stone in a tar pit – no splash, no ripples, no bubbles, no blushing. It's frustrating indeed for the innovator.

Somehow, we have to hold onto our faith and vision in new ideas, and be more willing to take risks – sometimes in spite of what the numbers say.

intelligence. As a society, we no longer suffer from the technological hubris that swept the nation during the halcyon days of the space program. We are overcoming this cockiness. I think today we have a more mature attitude about our place in the global marketplace. Most importantly, we have made – and continue to make – significant changes in the way we do business, the way we manufacture our goods and process information, and the way we manage innovation. As we make the transition from a manufacturing to a knowledge-based economy, we are slowly but surely casting off the worn management philosophies and outdated technologies that we stayed with far too long.

## Managing the innovative process

As a country, we certainly have the money to be as innovative as we want to. And we have the need, the appropriate technologies and the people. The problem, we have discovered, lies in how we have managed the innovative process within our large corporations. Interestingly, according to Peter Drucker, the U.S. is still the center for innovation activity in the world. But this activity today resides within the smaller entrepreneurial companies. As Drucker pointed out in a recent article, "over the last three years *Fortune* 500 companies have lost three million jobs, while companies under ten years of age have created more than one million jobs."

I see a lesson here. Many larger corporations in the U.S. are encouraging the development of pockets of entrepreneurialism within their organizations. Such as environment does encourage innovation. But small points of entrepreneurialism won't solve our bigger problems. We must also get our big institutions moving as well, if we are going to address the larger issues facing our society. You can't feed a country, and fulfill its energy requirements out of a garage-shop operation, financed by a couple of maverick engineers who borrowed $10,000 from their parents to build a microcomponent for a microcomputer. It still takes macro thinking, macro dollars, and macro organizations to keep western society moving forward.

So how do we do it? How do we inculcate innovation within one's own organization and infuse it into others? What is the magical process by which we transfer the innovative process across organizations and industries – and breathe new life into our aging industrialized societies? How do we make old managers think new again?

## An eight point plan

Let me suggest an eight-point plan for restoring the innovative process, and encouraging its replication in all types of organizations.

***First,*** we need to make innovation a responsibility of top management. Innovation should not be the sole responsibility of the research department. In the final analysis, innovation is not a new product or new invention, but a new attitude – one that must emanate from the executive suite and pervade the halls of factories of our companies.

***Second,*** we need to make top and middle management people accountable for innovation in our organizations. We need yardsticks that

drag on the wheels of change in many large organizations. He must know how to deal with the nay-sayers.

He must also overcome the negative economic and often sociological impact that invariably accompanies the introduction of any brave new idea. Robotics offers a classic example of this. The industrial robot had to survive a very difficult childhood – fraught with financial risk, union concerns over their effect on employment and job classifications and public misconceptions. The real heroes in the early days of the industry were not the robot manufacturers, but the first customers – the top management people at General Motors and other companies that championed the use of robots in their factories.

But the most admirable and important trait of our champion of innovation is his *attitude* toward innovation – and his ability to nurture it in others. In the entire innovation transfer process, I believe cultivation of the proper attitude may be the single most important criterion. A new idea will never succeed in an organization where innovation is regarded as merely a motherhood concept, or as a textbook theory. Innovation must be ingrained in the corporate culture, and become a way of life. Management must eat and breathe innovation. They must be given the leeway to embrace the entrepreneurial approach within their organizations without being penalized for any short-term negative impact on the bottom line. Instead, they should be measured on how well they transfer and cultivate the innovative process.

**Is it best to be second?**
In the United States, we still have many companies with a downright cynical attitude toward innovation. They have a curious R & D philosophy. Their strategy is to be the first company to be second. Wait until someone else has broken his pick pioneering, then jump in with vast resources to capitalise on the demonstrated opportunity.

Actually, from a bottom-line perspective, this attitude makes good sense. Look at all the successful imitators of IBM. They are like parasites that feed off their host. That works well – as long as the host stays healthy.

So why take the risk of being first, when the second bite of the apple tastes as good, if not better, and you know it is safe to eat? Why be innovative when you can develop a much better product through reverse engineering?

The problem with this attitude is obvious. It can all too easily spread like a disease and kill the total life system of an industry. Americans know this story all too well. Our steel, automative and machine tool industries are victims of this attitude. No one wanted to be first – except the Japanese. U.S. management and labor alike thought short-term, while the Japanese looked to the future. Our industry leaders preached profitability, while Japanese management championed the cause of innovation.

But fortunately, U.S. industry has turned the corner on innovation. We have been able to restore the innovation attitude in many of our industries, and have taken the lead in important new technologies, such as artificial

industry was adequately financed, and commercial development took off. Today, I think the industry is suffering from financial overkill.

The journey from concept to wide-scale commercialization of robotics took over 50 years. Actually, more like seven centuries, if you count the Medievalists.

There's been enough written about the short-sightedness of the Western businessman. I don't need to belabor the point. We have all heard the gospel of innovation according to the Japanese. But few Western companies have enough faith in the gospel to practice it in their daily business lives. And, believe me, it takes plenty of good old-fashioned faith to back an idea for ten to twenty five years, which is the average time span for a major technological innovation.

I've talked to Joe Engleberger, often referred to as the "father of the industrial robot", about the early days of the robotics industry. In his long list of patents and accomplishments, he is most proud of the fact that every time he went "back to the wall" for more financing, he got it. Conviction backed with financing – that's the formula.

Unimation Incorporated, now a wholly owned subsidiary of Westinghouse Electric Corporation, installed its first production robot in 1961. It made its first profit in 1975. The literature of innovation is filled with other illustrations of this innovation time lapse. Polaroid cameras, Xerography, piggy-back railroads have gone through this same agonizing cycle.

But you'll find that behind market demand, appropriate technologies, financing, and timing, there emerges a fifth ingredient for innovation – one that takes into account the *human factor* in the formula – the sweat, toil and conviction, the sleepless nights and the organizational politicking that are the very backbone of the innovative process.

**The Need for a Champion**

One theme that emerges from several studies on the innovative process done within Westinghouse is that behind every successful technology is a *champion* of that technology – one person who crusades for its development. This person is rarely the technical expert. rather, he is most likely a manager who backs the concept with unswerving conviction. More importantly, he knows how to lobby for its support within his organization and industry.

While this champion is normally an individual, it can be a group of people within an industry. In the U.S., a number of important innovations in the area of automation and electronics have been spearheaded by the military. The numerical control industry is a good example. NC machines have been around as a concept since about 1955. But there never would have been numerical control had it been left to the machine tool industries. It was forced upon them by our Air Force.

In addition to being a zealous missionary, the champion of innovation must also be a part psychologist and part sociologist. For the innovation process to succeed, he must know how to cut through the inertia that puts a

organization and infusing it into others. What industry needs is a practical "how to" manual on the successful transfer of the innovative process from one organization to another.

Let me offer a few personal conclusions derived mainly from hard lessons learned in the marketplace.

Ironically, most of what we know about innovation is based on hindsight, rather than on any innovating vision. One of the best studies on the topic was issued by the U.S. Air Force in 1967 entitled, appropriately enough, *Project Hindsight*. The Air Force decided that if they looked back and found what had succeeded and what had failed, they might be able to pinpoint the attributes that are necessary for the success of innovation. The study reached a three-point conclusion:

- The first requisite for successful innovation, the authors argued, was a perceived need. A basic point. But notice that the accent here is not on the need itself, but the *perception* of the need.
- The second ingredient is an appropriate technology to address that need, plus the knowledgeable practitioners of that technology.
- The final factor is adequate financing. If you miss any one of these three ingredients suggested by the Air Force, your innovation is doomed to failure. When all are present, you have the opportunity for success. But still, success is not assured.

*Timing* is that all-important fourth criterion.

### Innovation in the Robotics Industry

The robotics industry represents a classic example of the importance of timing to this innovation process. The concept of robotics has fascinated mankind since the middle ages. Legend has it that the medieval inventor, Albertus Magnus, invented a mechanical "android" in the year 1250 that could answer his door and perform the duties of a butler. The French clockmakers of the 1700's built several ingenious automatons that could write and draw like humans. But these devices were, of course, mere mechanical novelties – playthings for the aristocracy.

The real *need* for such machines never materialized until the 1900's, when the industrial revolution forced laborers and even children to perform machinelike tasks. Charlie Chaplin showed us what the industrialization of society was doing to the human psyche in the 1936 movie *Modern Times*.

Unfortunately, no *appropriate technology* was available at that time to rescue man from his enslavement to machines. Near the end of World War II, however, bits and pieces of the technology began to emerge. The big breakthrough came with the advent of solid-state electronics and digital logic. Appropriately trained people also were then available, coming mostly from industrial controls, nuclear energy, electronics and aerospace disciplines.

But it wasn't until 1956 that the industry was able to scrape together enough *financing* to survive. In the sixties and seventies, the robotics

# Cultivating Innovation within the Corporate Environment

## by Chester A. Sadlow
*Executive Vice President, Advanced Production Technology*
*Westinghouse Electric Corporation*

> ***Westinghouse Electric Corporation*** **is engaged principally in the manufacture, sales and service of equipment and components for the generation, transmission, distribution, utilization and control of electricity and in defense and high technology areas, such as radar equipment and robotics. Its businesses also include a wide range of products and services, such as broadcasting and cable television operations, community development, bottling and distribution of products, transport refrigeration and financing services.**

The word innovation summarizes in a nutshell the challenge facing American industry, and that of many other Western nations, today. Most people think this word means the introduction of a new idea or invention. Actually, the word comes from the Latin noun *innovatus,* which means to *re-*new, or make new *again.* That's the challenge we really face. How do we make our factories new *again?* How do we *renew* the innovative process that once characterized Western industry, but is now more generally associated with Eastern cultures?

Over the past five years, Westinghouse has been undergoing a rather extensive renewal process itself. We were once primarily a manufacturing company that served the electrical industry. While electrical products are still important to us, service businesses account for over half of our revenues. And over 70% of our payroll is made up of white-collar workers performing knowledge tasks. *Fortune* Magazine recently described the changes that have occurred within our company as a "cultural revolution".

So, as you can see, we have a keen interest in how change is managed in industry today. All of our top management people have gone to Japan to learn the Japanese approach to quality and productivity firsthand. We spend millions each year on management training programs. And we have done a number of internal studies on the innovation process within our corporate Research Center.

I have also read, with great interest, several recent studies on innovation – including the report prepared by Dr. Nuala Swords-Isherwood for the British-North American Committee[1]. This report does an outstanding job of compiling valuable data on the identification and nurturing of the innovation process. While such studies are indeed insightful, I think we need to take them one step further. What is lacking in the literature is a conceptual description of a *process* that can be universally replicated – that is, a process for inculcating innovation within one's own

---

1 Swords-Isherwood N. *The Process of Innovation,* British-North American Committee London, Washington, Toronto 1984.

this basic foundation, our proven ability to anticipate and apply technological developments would not exist.

**Conclusion**

I hope that my Northern Telecom "case study" illustrates that management plays a very crucial role in successful innovation.

There are, of course, many other aspects of the broad innovative process that I have not had time to address. These include optimum lab size; access to computers; the age ratio of R & D people; the relationship between R & D, manufacturing, and operations; and the rate of change of technology.

Each of these are indeed key factors that impact on an organization's ability to innovate – factors that deserve the committed attention of a firm's senior executives if they are to be managed for success.

Let me conclude simply by saying that no challenge is more vital to the success of our corporations, our industries, and our societies, than that of innovation.

strangers to BNR and the telecommunications industry. About a third of this group generally have already spent some time working for us, through summer placements or co-op programs.

Secondly – and most important – our influx of new researchers reflects more than the recent, absolute growth of our BNR facilities and the normal loss of individuals through attrition. Because of their technological expertise, our researchers are a vital source of talent – in such areas as marketing, sales, and general management – for Northern Telecom's operating subsidiaries, or for our mutual parent, Bell Canada, the telephone operating company.

As a result, each year between 3-4% of our BNR professional staff transfer over to the corporation or Bell Canada, a process we consider eminently desirable.

It offers our researchers the incentive of a broader career path, while ensuring that our operating subsidiaries contain a strong base of managers familiar with, and supportive of, R & D activity. Moreover, with the operating company experience they have gained, members of this group are positioned as excellent candidates to return to BNR as senior executives.

Managing a flow of talent to and through BNR, however, is only one way of ensuring enthusiasm and creativity. Another important solution to this challenge lies in what we call the BNR "capability fund."

To avoid stifling the creativity of the BNR staff, BNR is allowed to charge its Northern Telecom clients on the basis of cost plus 10% This 10% is the capability fund which BNR itself can devote to undertakings championed by its own people.

This means that a BNR researcher with a special interest or insight into a line of inquiry can apply to management for support. If he or she can convince management that this research might one day have some value to the corporation, resources are made available to undertake at least an initial study.

The results of BNR capability fund spending have included significant advances in fiber optics development, and in cellular radio systems.

**Design Standards**
There is one final "demand" I would like to raise that I believe must be addressed for innovative success.

- Designers work to a *world class standard* of excellence.

This means that we expect our designers to anticipate the products and technology our competition will be delivering, and then pursue designs that will leap-frog that competition.

We will *not* – because we dare not – settle for mere similarities or slight advantages. Such attributes are not sufficient to provide the comparative advantage we require to ensure growth and prosperity.

To achieve this goal, Northern Telecom and BNR have dedicated themselves, and spent the money needed, to develop significant expertise in all aspects of the core technologies of semiconductors and software. Without

Tax incentives are a better answer. They provide much greater freedom of action, and more positive results.

A *third* advantage of our approach to research and development is that, through the product line primes, we ensure that R & D is a *closely managed* activity, rooted in the needs of the marketplace, our broad corporate strategies, and – of great importance – our manufacturing capabilities.

Since the primes are also operating company officers, this helps guarantee that our researchers will take into account the constraints of manufacturing technology.

Product life cycles have shortened dramatically under the impact of new technology. In this environment, Northern Telecom cannot afford to waste *any* window of opportunity for sales, just because a new design needed reworking before it could leave the plant at the quality and cost we require.

This is why Northern Telecom and BNR have invested very significantly, and continue to do so, in developing and applying personal computers and the computer-assisted design and manufacturing tools that speed the design-to-manufacture cycle. We keep our labs up-to-date.

## People

Now let me turn to a third vital "demand" my corporation addresses to ensure successful innovation – people.

- This is the necessity to find structures and incentives that allow us to conserve our most vital R & D resource – the enthusiasm, expertise, and creativity of our researchers and scientists – our *people*.

I am sometimes asked how BNR "manages" the age of our R & D staff in order to ensure that necessary combination of experience and creativity. The first part of my answer is that we do not manage for age: rather, we manage for talent.

It is true, however, that we do pursue a steady flow of new researchers into BNR, currently equal to about 10% of our professional staff. Of this group, about half have just completed their undergraduate or graduate studies. As a result, the age of our BNR scientists and engineers averages in the early 30s.

Underlying this dynamic situation is the fact that Northern Telecom and BNR are acknowledged technological leaders in our industry. As a result, there does not exist an extensive outside pool of senior, experienced telecommunications researchers to meet our needs.

Instead, we believe in taking the brightest minds from the universities – generally, our new grads must be among the top 20% of their class – and then putting them to work under the guidance of the senior managers of BNR and Northern Telecom. It is this combination of our own seniors, who have an unparalleled depth of knowledge on the telecommunications network, with the commitment and vigor of our younger staff, that have delivered an industry-leading stream of successful new systems and products.

There are two other important aspects to our research staffing approach that I should mention. Firstly, our new-grad hires are often far from

In our view, R & D differs from capital investments, such as for plant and equipment, only in that it is fully expensed as incurred.

The research proposal must include a detailed analysis of the *market opportunities,* and the return it is expected to generate. And, of course, all R & D projects must be subjected to post-audits.

There are aspects to our approach that may be quite different from what you expect. For example, BNR *does not* carry responsibility for managing our R & D spending. It does not even receive a budget of its own in the normal sense.

Instead, our R & D spending is controlled through 'product primes', who are also senior managers in our operating subsidiaries. Each product prime is responsible for developing the global product portfolio recommendations in his area, such as switching or transmission. He buys his R & D wherever he wishes, or does it within his own organization.

Usually, for new products, he turns to Bell-Northern Research. It is generally the best equipped to provide the quality of people and range of expertise required for the initial innovation of systems.

From this you can see that BNR is not directly funded. Instead, it sells a design and development service to our operating companies.

This is a deliberate and direct incentive for BNR to hone its innovative skills and financial discipline.

Our matrix system of product primes, our investment-plan approach to R & D, and our incentive plans deliver three crucial ingredients that produce market-driven, profitable innovation.

***First,*** when an R & D plan involves a forecast of return-on-investment, it encourages the organization and its managers to look beyond the current, safely established product. In other words, it identifies the potential rewards that are the only true incentive for risk and change.

***Second,*** the annually reviewed investment plan provides a measure against which the project can be screened over time. Should the anticipated returns begin to decline – due to changing technology, shifting customer perceptions and markets, or problems in the research and development – we will have this early warning that the project should possibly be aborted.

This happens. And while disappointing, it is not all bad.

Well over half the R & D expenses of any new Northern telecom product occurs after the plant start-up. We see the investment-plan approach as helping us build a culture that allows abort decisions to be made as early as possible in the project cycle, so as not to waste valuable resources that could be better applied elsewhere.

Incidentally, it is this important freedom to abort speedily a project, when we decide that our investment could produce greater benefits elsewhere, that lies behind Northern Telecom's general lack of interest in R & D programs directly sponsored by government. While these can play a vital role for smaller firms that lack initial research resources, it is also true that the political agenda rarely permits the rapid, market-based decision-making a corporation such as Northern Telecom requires for successful innovation.

The heart of our R & D activity lies in a subsidiary, Bell-Northern Research, better known as BNR. It employs about 4,000 people.

In addition to BNR, Northern Telecom also supports some 27 R & D centers of its own, based in plants throughout Canada, the U.S., and the U.K. These employ a further 1,500, for a corporation-wide total of 5,500 people working on R & D.

**Culture**

R & D facilities and staff, however, no matter how large, are no guarantee of market success. Which brings me to Northern Telecom's "demands" for innovative performance.

- First, there is the demand for a corporate-wide *culture* with two linked attributes. This culture must regard innovation as a paramount virtue, and also define R & D in terms of the marketplace and the customer.

One of the most important roles the chief executive and his senior officers can play in any business is to insist on the value of innovation and change. I have found that, unless this demand is championed from the top, constantly and honestly, the natural inertia of any large, established organization can easily triumph.

Above all, at Northern Telecom, it is the responsibility of top senior management to make sure our R & D labs and centers have the financial and human resources that successful, profitable innovation itself demands. We play a full and active role in setting the total R & D spending levels, sorting out the priorities, and establishing the proposed development schedules.

At Northern Telecom, we do not view R & D funding as a tap to be turned on and off, depending on short-term profit objectives or economic constraints. To do so, would deliver a clear, tragic message to the culture. It would identify R & D, innovation, and change, as an option rather than a constant necessity.

Instead, we are committed – even in our rare lean year – to providing the maximum funding possible for research and development.

This year, our R & D investment is expected to equal about 10% of global consolidated revenues, compared to the 6.9% of much smaller global revenues we spent in 1978.

Yet the funding for R & D must take into account a temptation that can equally sabotage innovation.

Researchers and scientists, by their very nature, can be seduced by the desire to develop technology for its own sake. But product development which fails to address real market needs and user interests, is a waste of vital competitive resources.

**Investment**

How then does Northern Telecom resolve this dual demand – the need to encourage change, while ensuring it is market driven? The answer starts with our second corporate "demand" for managing innovation:

- Innovation must be regarded as, and managed as, an *investment*.

# The Demands of Innovation in Northern Telecom

## by Walter F. Light
*Chairman and Chief Executive Officer*
*Northern Telecom Limited*

> **Northern Telecom Limited** is the second largest designer and manufacturer of telecommunications equipment in North America and sixth in the world. It is the world's largest supplier of fully digital telecommunications systems, and is a significant supplier of integrated office systems. Revenues in 1983 exceeded $3.3 billion. It employs more than 43,000 people throughout the world, and has research and development facilities and 46 manufacturing plants in Canada, the United States, United Kingdom, Republic of Ireland, Malaysia and Brazil.

Business exists today in a global environment of dramatic, accelerating technological change, and of increasing economic competition. In this environment, successful innovation has become the most important demand placed on many companies.

One of the best definitions of innovation has been offered by Peter Drucker.

"Innovation," he wrote, "is not a term of the scientist or technologist. It is a term of the businessman. Innovation means the creation of new value and new satisfaction for the customer."

Those attributes – customer value and satisfaction – are the imperatives of corporate success. At Northern Telecom, we believe in them passionately. They are an integral part of our corporate culture.

I believe you will agree that, while scientific *invention* may occur by accident, industrial *innovation* never does.

But it is possible for a company to ensure – in effect, to *guarantee* – successful innovation shall take place.

It starts when senior management *demands* it must. Then, management must back this demand with hands-on commitment, organizational planning, market savvy, and the necessary resources, both financial and human.

Most of all, the demand for innovation by senior management must be framed in terms of customer needs and, equally important, customer perceptions.

For Northern Telecom, the heart of the innovative process lies in a continuing emphasis on *market-driven, profitable* R & D.

It was our strong R & D commitment and capability, and large system expertise, that allowed us to anticipate the impact of microelectronics and software on telecommunications technology, and to initiate development programs for a constant stream of digital telecommunications products and systems.

keeps a sense of curiosity about natural phenomena and many quickly develop the tenacity necessary to deal with the complex problems of industrial research.

Even the successes of our scientists will not be seen by them in the marketplace perhaps for many years and this is why they need the encouragement and support of the top management in the business. I believe that in Unilever we give them that support because we think their work is vital to the future of this business.

therefore to examine the implications of biotechnology for us. Let me mention two such areas to illustrate the way our fundamental research programme has led to innovations in biotechnology. Some time ago, we set up research teams in the following areas:

- *Immunology*, to develop skills and knowledge which we felt would help our animal feed business in combating diseases of pigs and cattle.
- *Controlled Cell Culture,* to develop techniques in culturing single plant cells to produce a mass of identical cells.

Later, these groups were supported by a small team working on genetic engineering in both bacteria and plants.

An important feature of these is that they are all under one direction and whilst working on quite fundamental topics, the groups have a fairly clear view as to where their work might be exploited.

So far, immunological work has produced the following innovations:

- an oral vaccine for pigs and calves which is used as part of animal feed and cuts down mortality and morbidity in these animals dramatically and
- medical diagnostic devices for humans, and they are now developing (with the aid of the cell culture team) the manufacture and use of monoclonal antibodies for early diagnosis of disease in humans.

The cell culture team has succeeded in producing clones of identical oil palms which are now on our plantations in Malaysia and which give a yield increase of over 30%. We have also established a company to market cloned palm seedlings to non-Unilever plantations throughout the world. Parallel work on the coconut palm is already well in hand.

**People**

In Unilever we have no magic formulae which we can use to show how efficient we are at research and innovation. The essential investment we make is in people; people who are skilled and creative and we must ensure that we can recruit bright young scientists. Where possible, we look for young men and women who have a proven record of research, mainly at universities and we encourage them to exercise their curiosity by making sure that they have considerable freedom to select their problems and to publish the results of their research, if necessary after patents have been applied for. We offer a wide spectrum of opportunities for such people to develop their careers within Unilever. Some will continue in the research division and make their mark in the world of science, some will go into technical development and others may opt for technical or even business management in Unilever companies.

We consider this mobility good for both our research function and the business. The business receives a flow of well-qualified and up-to-date technical people who understand the strengths and limitations of research and thereby use it the more effectively. For its part, research gains extra points of contact with the business but this mobility also enables it to recruit regularly and to ensure a continuing flow of good scientists through its laboratories. We encourage this flow of young people through research; it

There are many barriers to innovation, both social and intellectual. Society has a natural tendency to resist change; a tendency which is not only normal, but in some ways even desirable. An organisation totally devoid of resistance to change would fly apart at the seams. It *must* both seek it out and resist it.

The reality is that radical product innovation can mean radical changes in all phases of the business. It can mean new product techniques, new channels of distribution and perhaps even a new conception of the marketplace. Just think, for example, of the great advances in immunology which give unbelieved scope for early (even incipient) diagnosis of a wide variety of diseases. Ally this to advances in micro-electronics and the potential for home diagnostics is enormous. Over-the-counter diagnostic kits for many problems are only a few years away.

**Teamwork**
There is, in the innovative process a need to discipline and pull together in a co-ordinated way the many functions that are involved. Innovation – certainly on anything other than a small scale – is a matter of teamwork. It involves managing a team drawn from research, engineering, production, marketing, financial and other functions. Each of these functions will have its own way of viewing innovation and its own role in the process.

Each will have different backgrounds, motivations, and certainly different timescales. Even when each is well-intentioned there will be little innovation unless all of them understand clearly their role and work to a common purpose. The process involves transferring information, enthusiasm and authority from one group to another, and this always requires a change in the activity patterns of the various groups. As far as industry in concerned, it is a process that requires the most creative and competent management skills imaginable.

We at Unilever see plenty of opportunities for successful innovation in both products and new areas of activity. We expect that many of these will derive from the combination of development in the bio-sciences and engineering generally known as biotechnology.

**Biotechnology**
Micro-organisms and animal and plant cells are so versatile that through special fermentation techniques they could meet virtually all man's needs for organic chemicals. This is the basic concept of biotechnology and its importance for the future of the industry is that it will almost certainly allow new manufacturing industries to be set up that are based on resources renewable by photo-synthesis.

The key feature of most advances in biotechnology that we have made is that they have been based on multi-disciplinary approaches. Success in this field involved the co-operative activity of microbiologists, biochemists, chemists, physicists and engineers. We have had such multi-disciplinary teams working together for a long time supporting our existing food, detergents and personal product businesses and we were in a good position

## Timing

In the mid-fifties our research groups were trying to simulate a variety of meats by using first groundnut protein and later soya protein. They produced simulated ham, luncheon meat and chicken breast, which were nutritious and very realistic but they were far too early and even now the public is not quite ready to accept substitutes at worthwhile prices.

The second example is related to the cost of energy. After some very good scientific work we developed in the early sixties a formulation that enabled detergent powders to bleach clothes at medium temperatures instead of at the boil. The saving of energy is clearly evident but the invention was too early and it took the OPEC price rises to make the public aware of the cost of energy before we were able successfully to launch new products based on this development.

On the other hand, the work of our biochemists to establish the beneficial health effects of poly-unsaturated fats, and the skills of our development scientists in formulated margarines containing considerable amounts of such fats, were exploited very quickly.

The most interesting aspect of the first two examples is that both were in product areas where we have a great deal of experience and yet our timing was awry. Imagine how much more difficult it is to judge those new inventions in a field remote from our established groups. And yet, we must always try to assess their value to us and to exploit them in spite of the risks when we believe that they have a profitable future. But most important of all, we have to refine our ability to estimate the time factor and so ensure speedier and more efficient commercial exploitation of new developments. The risks of innovation are high and I suspect that we shall make some mistakes of timing in the future but the rewards are also likely to be high and we must constantly seek for such rewards. In any case, change is endemic in business and we must embrace it and where possible direct it to our advantage.


## Markets

The really important point is that of the market dimension. The concept of an invention seeking a market, although it occasionally occurs, is one which is not attractive nor, on the whole, productive. I believe that we have to be able to determine the needs of the market place before we can hope to match them with the appropriate technology. In other words, an important element in innovation is the analysis derived from the market research of the needs of the consumer with all that that implies in terms of price, design and so on. In 1963 a book called "Britain 1984" was written by an economist, Ronald Brech, as an attempt to forecast what life would be like in the year already made famous by George Orwell. The book made no mention of stereo decks, VCRs, micro-computers of cable TV – even though all these were technically possible at the time. In 1984 more than 2 million personal computers will be sold in the U.K. in a market that did not exist five years ago.

# Innovation – Its Importance and its Problems

## by Kenneth Durham,
*Chairman of Unilever plc.*

> *Unilever,* with its headquarters in London and Rotterdam, provides a wide range of locally-manufactured products and services in more than 70 countries. Whilst the larger part of the business is in branded and packaged consumer goods (mainly foods, detergents and toilet preparations), other important activities include chemicals; paper, plastics and packaging; animal feeds; transport; tropical plantations; and the diverse operations of UAC International, one of its largest subsidiaries.

We often hear that in the United Kingdom we are brilliant at fundamental research but not so good at translating this into processes and products. A recent survey by Japan's Ministry of International Trade and Industry (MITI) showed that since the Second World War the U.K. has produced 55% of the world's "significant" inventions. (The United States has produced 22%, France 14% and Japan 6%).

We are, it is said, an unenterprising nation. And we are told, this time I think with justification, that one of the reasons for our lack-lustre domestic economic performance is our dilatory response to technical change. Whatever the causes, one thing is indisputable: with one or two exceptions, our industry has not kept pace with its international competitors in its ability to innovate.

Nevertheless, I feel sure that we *do* have the capability, given the right environment, to exploit new technology (and certainly to develop it), and through that capability, to produce new and unique products competitively.

In Unilever we believe strongly that product innovation is essential for the future health of the organisation and we think that advances in science and technology are likely to be the most important source of new ideas. Innovation can mean quite radical changes in many aspects of the business and we in leadership positions must endeavour to give encouragement to, and support for, the generation and development of new ideas.

There is sometimes a misconception that invention and innovation are essentially the same. We do not take this view; we consider that innovation is by definition the commercial exploitation of a new idea from whatever source. In our experience innovation has three dimensions. It is the synthesis of a particular consumer need and the technical means of satisfying that need. But is has one other dimension, that of time; the synthesis must take place at a time when the market need is clearly expressing itself. Some examples from Unilever Research will illustrate the point.

pool of genetic material. So an individual with a special interest was hired to perfect the process for our use and we see that this will help our breeders develop better hybrid because hybrid vigor results from the crossing of unrelated lines. In addition, plant breeders saw this as a means for our production people to measure quickly the purity of our seed production. Rather than having time consuming and costly growouts, we now have quality control laboratories equipped with electrophoresis equipment. There are six locations for doing this type of work including two in Europe (one in France and one in Austria). It is possible to test the seed for purity about three weeks after pollination. This would be about five weeks before harvest. Normal growouts would require ten weeks after harvest and would not be as accurate.

Although Pioneer's success in hybrid seed corn can be attributed to a high degree to its research efforts, it also has very successful production and marketing organizations, both of which have been quite innovative in their own right.

## Production plants

Pioneer has the most modern efficient production plants in the country. This is due in part to the fact that it started several years ago with in-house construction management services but the user of the service made the final decision about design rather than vice-versa. Throughout the company we work very hard to see that the people affected by a decision or plan provide the input into that decision or plan. We advocate participative management and find that employees have a very good ability for being innovative given the right environment and encouragement.

One such production innovation is what we call our dryer bin prediction system. Seed corn is harvested on the cob at an average moisture content of 30% and it is very slowly dried to 12% before shelling. The process takes about three days on the average but can range from 2-4 days depending upon initial moisture content and the particular hybrid. Until recently we sampled and tested the seed for moisture every few hours to make sure it was drying properly and that the drying was stopped at the right time. Too much drying wasted energy while too little could cause germination and vigor problems. Now, through electronic monitoring of intake and exhaust temperatures and a central recording system, we can accurately predict with a programmable calculator the exact hour to shell each bin without ever going into the bin. This system was developed with the help of an agricultural engineering professor who was hired by a divisional production manager as a consultant on drying. There was one employee who embraced the concept and worked with the consultant to see the system develop into a workable one. Enhancements continue to be made to the system reducing the amount of data that must be put manually into the system.

## Marketing

Currently there is an innovation going on in marketing that is unprecedented in the history of the company. We have in the field with our sales agents

1,500 portable computers. By the end of this year, 3,000 of the 5,000 sales reps in North America will be equipped with the portable device. The unit serves for order taking, inventory control and invoicing. Information is transferred to our central data processing via a phone modum. Theoretically, at least, each morning the field activity of the previous day is known. The whole system is the result of years of work of:
- trying to cut down on the volume of paper handled
- the desire to speed up information flow, and
- an effort to reduce errors in the order and entry and invoicing process.

Most of the credit for this program goes to the marketing people in one division who saw the opportunities for such a system. Having our own computer company that specialized in portable equipment was also a stimulus!

When I visited with various people in the organization to discuss innovative ideas and the process, there were several things that were common:
- The stimulus was often the need to solve a specific problem
- Ideas generally came from group discussions of the specific problem
- Someone accepted the advocacy role for an idea even though the idea often came from someone else
- Work atmosphere encouraged innovation.

What is meant by "work atmosphere encouraged innovation"? Pioneer has a very decentralized organization with several divisions that operate quite autonomously. In North America there are five operating divisions in the U.S. and one in Canada. Twenty two production plants are involved in seed corn production. These operating units are small enough for most employees to realize that they can impact the decision making process. Participative style management is encouraged right down to the individual work units at individual locations. New ideas and experimentation are encouraged. We are working hard at trying to see that supervisors at all levels are adequately trained in skills required, but especially how to go about asking questions that provide the stimulus for ideas that can increase productivity.

**Service support**

The divisions that produce and market our products are provided with service support at the corporate level. In addition to product development, service and support are given in the areas of electronic data processing (EDP), legal, construction and engineering, human resources and office services. With few exceptions, divisions are free to use or not to use services offered at the corporate level. This system works well for us. The service people must sell their service and the divisions that produce and market our products are very clearly competing for performance. This encourages innovation and makes it possible to test new ideas that would otherwise go unnoticed in a centrally structured organization. For example, we have just changed to a new sizing system for corn seeds because one division had an idea they felt would significantly improve the older system. That division adopted the new system and proved that it would work well. If all divisions had to agree prior to a change this innovation would not have come about, at least not for some time.

In conclusion I would suggest that, despite the fact that we are considered in our industry to be quite innovative, and I believe we are, there is still a great deal of room for improvement.

# Innovation within Pioneer Hi-Bred International, Inc.

## by Carrol D. Bolen
Vice-President, Pioneer Hi-Bred International Inc.

> *Pioneer Hi-Bred International, Inc.* is the leading supplier of hybrid seed corn in the world market. The company also produces and distributes other seeds, inoculants and computers. Located in Des Moines, Iowa, the company employs 2,700 people.

Pioneer Hi-bred International is in what is known as the genetic supply industry. Pioneer is on the input side of agriculture. The primary business is to supply farmers with seed stocks to produce crops.

Pioneer was founded in 1926 by Henry A. Wallace and some friends to market hybrid seed corn. Henry Wallace was the U.S Secretary of Agriculture for eight years in the 1930s and served one term as U.S. Vice-President from 1941-45. He had been experimenting with hybrid maize for a number of years and when he had developed a hybrid that he thought might be of commercial benefit to farmers, he and his friends formed the company in 1926 to produce and market the products. Until that time hybrid corn was considered to be a scientific curiosity but not commercially practical.

Hybrid seed was expensive and few people believed that farmers could be convinced to pay the high price required. It was a struggle in the early years and required an innovative marketing program. A few salespeople were hired to go into targeted areas and give samples of hybrid seed to farmers to compare against their open pollinated varieties. Farmers who accepted the new hybrids were then encouraged to become part-time sales representatives selling to neighbours on a straight commission basis. There were two major reasons for this selling concept. First, the company could not afford to pay the number of salaried salesmen that would be required. Second, farmers would be more inclined to buy from someone they knew and who was using the product himself. So the company was founded by an innovative individual, with an innovative product through an innovative marketing organization.

Although we now research, produce and market several products other than seed corn, this paper will be limited to our first and still major product, hybrid seed corn.

## Research
Since the company was founded by a researcher, research has played an important role in the company since the very beginning. That commitment to research has enabled it to continue to grow over the years until it now has 36% of the U.S. seed corn market, three times that of its closest competitor, and a sizeable share of the market outside the U.S.

I recently visited with our plant breeding management team about why that program has been so successful. Although not in any particular order of importance, the items discussed were:
- Adequate funding
- Mechanization and automation, including information processing
- Decentralization and individual freedom but with a high degree of communication between individuals
- Use of task force for planning and problem solving.

In the corn research group the company has 48 scientists plus technician support working at 24 locations in North America. In addition they are supported at the corporate level by 9 people in administration, 8 plant specialists and 7 data management specialists.

Pioneer Hi-Bred's program is referred to by management in corn breeding as "fast break corn breeding" meaning the ability to turn out improved hybrids quickly. You may question use of the word "quickly" because on the average it takes about eight years from initial development work until a new hybrid is commercially available. While developing new hybrids is not difficult, developing superior ones is very difficult. Although it is a gross over-simplification, one can say that success in plant breeding is a numbers game – the more products you develop the more apt you are to have successful ones. And in the last seven years Pioneer's corn breeders have increased their testing by 50% with only 4% more people.

Each year one of the major goals is to see how more research can be conducted per employee. Individuals talk about the restraints and possible solutions are investigated. We have in recent years moved from a hand planted, hand harvested, hand weighed, paper recorded system to mechanical planting and electronic data collection at the source. Most equipment is either custom designed or significantly modified, commercially available equipment. Mechanical equipment is designed by an in-house engineer and electronic data collection and processing is handled by our own data systems division in co-operation with the plant breeding division's own data management group.

**Technologies**
We consider ourselves a "high tech" company. Although very little basic research is done, we do take new technologies and adapt them to our operations. The goal is to do this better and faster than our competitors. These new technologies come from a wide array of sources but we have, in plant breeding over the years, relied heavily upon the public sector for innovative ideas. One such innovation is a process called isoenzyme electrophoresis that enables a person working in a laboratory to determine quite accurately the similarity of different genetic materials through a fingerprinting process that measures enzyme differences by using a special lab gel and electrical charge. It was used on fish species in the 1950s and then on plants in the 1970s. One of our breeders saw the potential for electrophoresis to help identify genetic differences or similarities in our huge

# Preface by the Joint Chairmen

The British-North American Committee has recently concerned itself with important elements in our three countries' response to increasing global competition. One of these is industrial innovation.

In 1981 we issued *Industrial Innovation in the United Kingdom, Canada and the United States* by Dr. Kerry Schott, and in 1984 we are publishing *The Process of Innovation* by Dr. Nuala Swords-Isherwood.

In addition to sponsoring these publications, at a meeting of the Committee in Torquay, England in June 1984, we asked several members to present brief reports on their personal conclusions to the question, "What Makes Innovation Work?" This publication contains their contributions. We consider that the experience and opinions they contain would be useful to a wider public. We are pleased therefore, to make them available in an Occasional Paper to coincide with publication of the Report on *The Process of Innovation*. The latter appears among our regular series of studies and is based on interviews with thirty six innovative firms in the three countries represented.

A. B. MARSHALL
*Co-Chairman*

WILLIAM L. WEARLY
*Co-Chairman*

WILLIAM I. M. TURNER
*Chairman, Executive Committee*

# Contents

The British-North American Committee .................. INSIDE FRONT COVER

**MANAGING PRODUCT INNOVATION**
*An Occasional Paper by*
*Carrol Bolen, Ken Durham, Walter Light,*
*Chester Sadlow and Viscount Weir*

**Preface** ............................................................................................ v

**Innovation within Pioneer Hi-Bred International Inc** ....................... 1
by Carrol D. Bolen,
*Vice President Pioneer Hi-Bred International Inc*

**Innovation – Its Importance and its Problems** ................................ 7
by Kenneth Durham,
*Chairman of Unilever plc*

**The Demands of Innovation in Northern Telecom** .......................... 13
by Walter F. Light,
*Chairman and Chief Executive Officer, Northern Telecom Limited*

**Cultivating Innovation within the Corporate Environment** ............. 19
by Chester A. Sadlow,
*Executive Vice President, Advanced Production Technology,*
*Westinghouse Electric Corporation*

**An Innovation of the Weir Group** .................................................. 27
by Lord Weir,
*Chairman of the Weir Group plc*

**Members of the British-North American Committee** ...................... 33

**Sponsoring Organisations** ............................................................. 37

**Publications of the British-North American Committee** ................. 39

© British-North American Committee 1984
Quotations with appropriate credit permissible

ISBN 0 902594 44 3

Published by the British-North American Committee
Printed and bound in United Kingdom
by Contemprint Limited, London SE1

October, 1984